A PLANET IN PERIL

ATTACK OF THE TOXLINGS

WRITTEN BY ROBIN TWIDDY

BookLife
PUBLISHING

©2023
BookLife Publishing Ltd.
King's Lynn, Norfolk
PE30 4LS, UK

All rights reserved.
Printed in Poland.

A catalogue record for this book is available from the British Library.

ISBN: 978-1-80155-890-7

Written by:
Robin Twiddy

Edited by:
William Anthony

Designed by:
Amy Li &
Drue Rintoul

All facts, statistics, web addresses and URLs in this book were verified as valid and accurate at time of writing. No responsibility for any changes to external websites or references can be accepted by either the author or publisher.

AN INTRODUCTION TO BOOKLIFE RAPID READERS...

Packed full of gripping topics and twisted tales, BookLife Rapid Readers are perfect for older children looking to propel their reading up to top speed. With three levels based on our planet's fastest animals, children will be able to find the perfect point from which to accelerate their reading journey. From the spooky to the silly, these roaring reads will turn every child at every reading level into a prolific page-turner!

CHEETAH
The fastest animals on land, cheetahs will be taking their first strides as they race to top speed.

MARLIN
The fastest animals under water, marlins will be blasting through their journey.

FALCON
The fastest animals in the air, falcons will be flying at top speed as they tear through the skies.

Photo Credits

All images are courtesy of Shutterstock.com, unless otherwise specified. With thanks to Getty Images, Thinkstock Photo and iStockphoto. Front Cover – calvindexter, Mykola Mazuryk, Loco, shufilm. Recurring – Pete Pahham (Angelina), Galyna Lysenko (green smoke), MG 5G, Somchai Siriwanarangson, Yellow Cat (toxlings). 4–5 – Design Projects, Limbitech. 6–7 – higyou. 8–9 – Roengrit Kongmuang, arleksey, sdecoret, ktsdesign. 10–11 – Everett Collection, sdecoret. 12–13 – CalypsoArt, Gorodenkoff, Eugenio Marongiu, Bugtiger. 14–15 – CDC (wiki commons), Grindstone Media Group, Tatiana Chekryzhova, Ultimate Travel Photos. 16–17 – dwphotos, Jarun Ontakrai, Morten B. 18–19 – Alliance Images, Dragon Images, New Africa. 20–21 – D_Townsend, Daisy Daisy, UGstockstudio, Andrei Kuzmik, Peyker. 22–23 – andrey_l, Kirill Gorshkou, noomcpk, nik sriwattanakul. 24–25 – Eugeniy Marin. 26–27 – pcruciatti, Gorodenkoff. 28–29 – Vladimir Borovic, Prostock-studio, Eviart, ktsdesign.

CONTENTS

- Page 4 — Far above the Earth
- Page 6 — Meanwhile
- Page 8 — Abducted
- Page 10 — Changing the Planet
- Page 12 — Men in Black
- Page 14 — Plastic for Everyone
- Page 15 — Pesticide
- Page 16 — Toxic Toys
- Page 18 — Cleaning Can Kill
- Page 20 — Poison in the Walls
- Page 22 — The Toxic Town
- Page 24 — Evidence
- Page 26 — Saving the Day
- Page 28 — The Big Clean-Up
- Page 30 — Eyes on the Skies
- Page 31 — Glossary
- Page 32 — Index

WORDS THAT LOOK LIKE THIS ARE EXPLAINED IN THE GLOSSARY ON PAGE 31.

FAR ABOVE THE EARTH

High above Earth, a plot was being revealed. An alien <u>species</u> with plans for Earth was getting ready to bring these plans to their conclusion.

Finally, after years of guiding humankind from the shadows... pulling their strings with our slimy tentacles... Earth is almost ready for the taking.

> They said we couldn't convince humanity to poison their own world. **Ha ha ha ha!** Look who is laughing now. **Ha ha ha ha!**

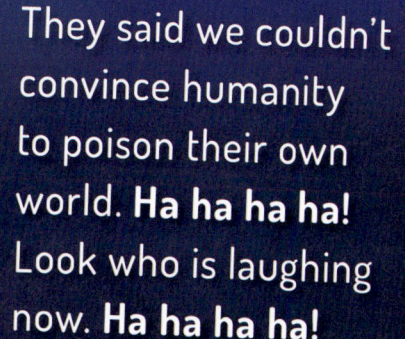

Fraxulon initiated the second-stage turbine and the strange metallic vehicle zoomed at an impossible speed towards the Earth. Breaking through the <u>atmosphere</u>, it headed towards a small town in the UK.

MEANWHILE

Back on Earth, Angelina Buttersworth was collecting batteries to recycle. Suddenly she saw something strange in the sky!

Angelina watched the strange craft from the bushes, as it landed in a nearby field. Two odd-looking creatures floated out.

The strange red creature said, "Release the Tri-mono Monitor." A hovering box came out of the spaceship.

The box scanned the grass and air. It took samples of the soil and water in the park.

"The Tri-mono Monitor has finished scanning," said the blue-tentacled alien.

Angelina tried to be brave. She tried really hard. She tried not to scream. But she did scream. She screamed very loudly.

ABDUCTED

Angelina grabbed her bike, but the aliens were quicker. Soon, she was pulled into the air and onto the ship.

Angelina was trapped in an energy field. "Who are you? What was that box?" she asked.

"We are the Toxlings from the planet Toxonia. It is a beautiful planet filled with <u>toxic</u> swamps and lovely acid rain," said the red thing.

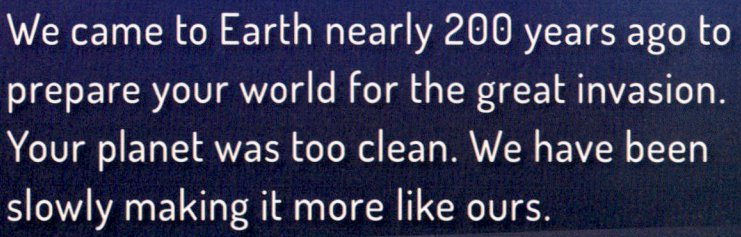

We came to Earth nearly 200 years ago to prepare your world for the great invasion. Your planet was too clean. We have been slowly making it more like ours.

We are so close to completing our mission. We can't let you go. It didn't take long to get you stupid humans polluting your own planet.

CHANGING THE PLANET

The human race was just starting to build factories when we first arrived. They called it the industrial revolution. We called it our way in.

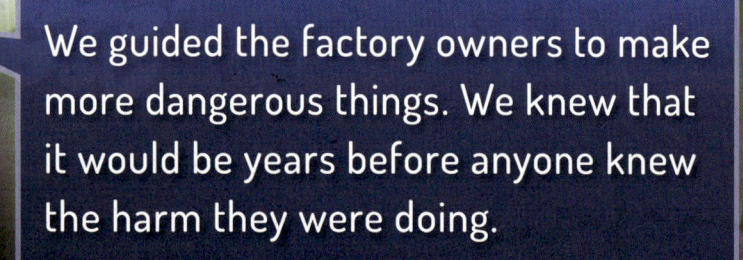

We guided the factory owners to make more dangerous things. We knew that it would be years before anyone knew the harm they were doing.

PEOPLE AREN'T STUPID. THEY WOULDN'T LISTEN TO YOU TWO.

It was simple. We left the right things where the right people would find them. A <u>formula</u> here, a piece of plastic there.

We did lots of our best work during and after World War Two. Both sides would have done anything to win. We were happy to help.

New ways to make things meant new ways to pollute.

MEN IN BLACK

In the 1950s and 1960s, there were lots of sightings of people who looked almost human. These were our 'Men in Black'. They joined governments around the world.

Plastics are not found in nature. They often do not break down in the same way that natural materials do. Many actually release toxins as they break down.

PLASTIC FOR EVERYONE

Think about how much plastic you see and touch every day. I bet it is a lot. Plastic is perfect. Making it makes toxic waste. Melting it makes toxic waste.

The plastic plan worked better than we ever thought it would. It can be found everywhere now: the oceans, the land, inside animals and even in the human body.

PESTICIDE

It was really helpful for us when you started covering your crops in poison to get rid of pests.

Don't forget DDT. That was a very harmful <u>pesticide</u> used in America. It is now over 40 years since it was banned and we can still find traces of it in vegetables and even people.

TOXIC TOYS

Some of the most toxic things on the planet are <u>heavy metals</u>. You stupid humans are putting them in everything.

Angelina wondered if it was the same sort of heavy metal that her dad listened to. She hoped not. She didn't like that very much.

Lots of electronic devices have parts that are made from heavy metals, such as lead, cadmium and mercury.

What do you think happens to your electronic waste when you throw it out? It goes to landfills or sent to other countries. Then, those toxic metals slowly seep into the environment.

CLEANING CAN KILL

Have you ever wondered why so many cleaning products say things such as, 'hazardous to humans?' We did that.

The average home has around 62 toxic chemicals in it. What clever villains the Toxlings are. They made being clean toxic.

Cleaning products can pollute the human body when they are used. They also pollute the environment when they are washed away or thrown out.

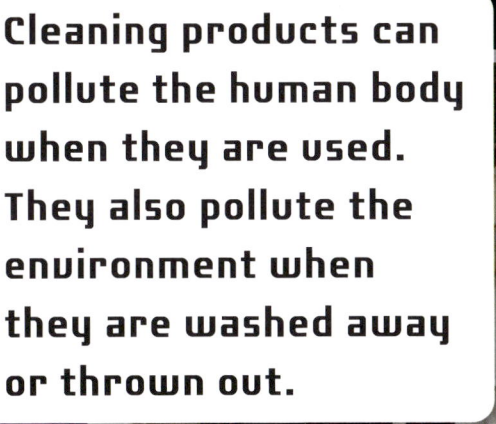

Most of these chemicals are harmless in small amounts. However, they can be stored in the body.

I pride myself on my ability to store toxins in my body. I can release them as a gas!

POISON IN THE WALLS

We managed to convince humans that asbestos was a miracle product. It is a material made from lots of little fibres. The dust from it is very bad for humans, but it is very good for Toxlings.

Humans did eventually work this out and stopped using asbestos. It is such a shame. We Toxlings really like asbestos.

What about lead paint? That was a good one.

Lead is very toxic for humans and the environment. We managed to get humans to paint their walls with it.

It got banned eventually. Lead is a tasty treat on our planet. It doesn't seem to be as popular here.

THE TOXIC TOWN

Manufacturing can create toxic waste as a **by-product**. One company decided to move their toxic waste on the backs of open top trucks through a town called Corby.

Those messy humans spilled toxic waste all over the town. Thousands of those barrels were lost or dumped in ditches and trenches.

Corby is still considered a great holiday site for Toxlings.

Even better is The Love Canal. It is a small town in the US near Niagara Falls. It was built on a chemical dumping ground.

Toxic sludge started bubbling up out of the ground. It was found in playgrounds, schools, homes and streets.

EVIDENCE

If the Toxlings had been less confident, they might have checked Angelina for electronic devices. If they had, they would have noticed that she had recorded their whole <u>confession</u> on her phone.

> YOU MONSTERS! I AM GOING TO TELL EVERYONE ABOUT YOUR PLAN. THEN, WE WILL CLEAN THE PLANET UP AND KICK YOUR SLIMY TENTACLES OFF IT.

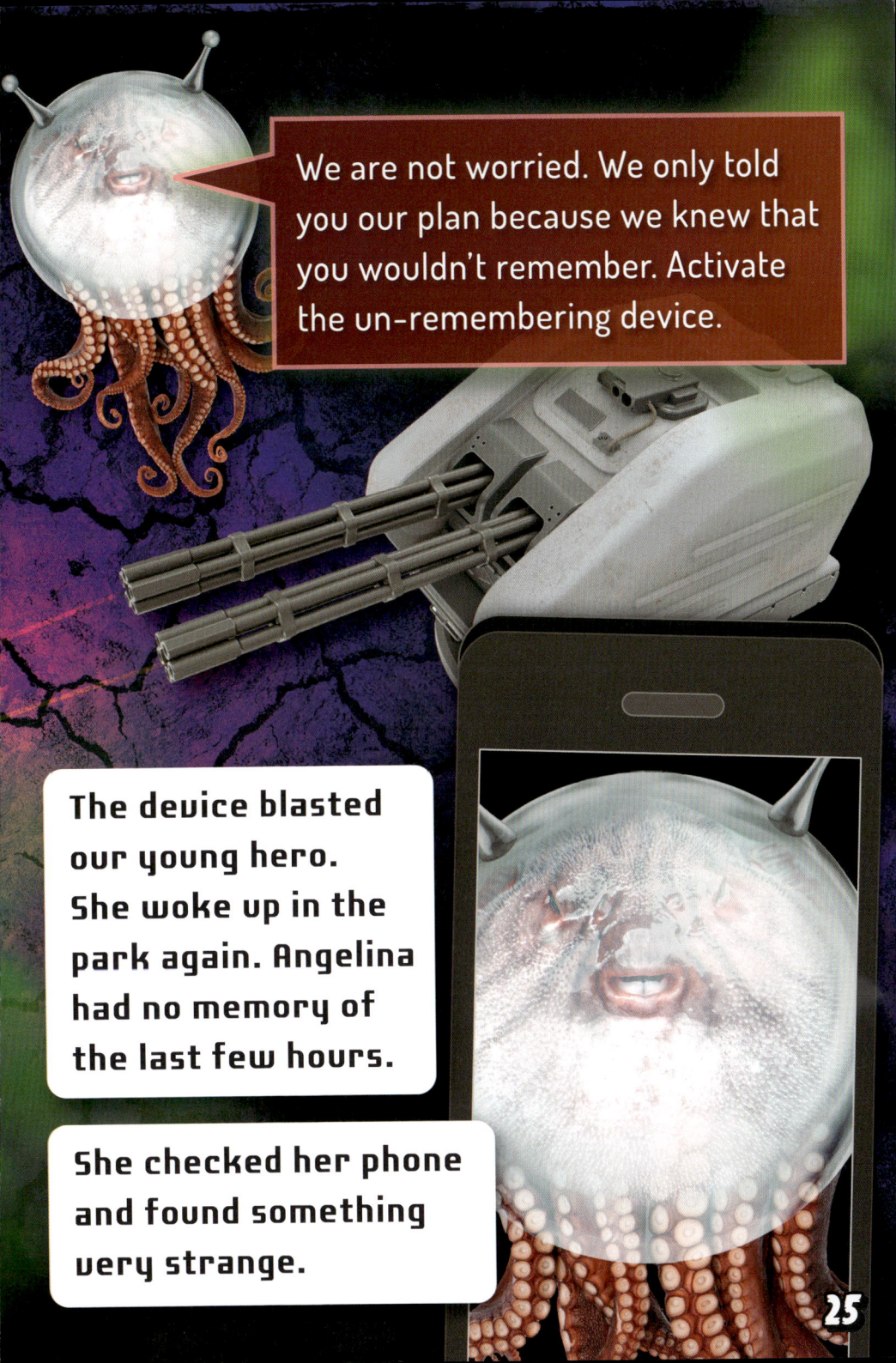

We are not worried. We only told you our plan because we knew that you wouldn't remember. Activate the un-remembering device.

The device blasted our young hero. She woke up in the park again. Angelina had no memory of the last few hours.

She checked her phone and found something very strange.

SAVING THE DAY

With the recording in hand, Angelina ran straight to Prime Minister. She showed the Prime Minister the recording and explained that something needed to be done.

The news was interesting that evening.

Breaking news. A young girl has uncovered a global conspiracy involving world leaders, heads of industry and aliens.

NEWS

Angelina was asked to hold a press conference. She told everyone what had happened and what needed to be done next.

WE MUST ALL TAKE RESPONSIBILITY FOR THE WORLD WE LIVE IN. WE MUST ALL WORK TOGETHER TO MAKE THIS WORLD CLEAN AND NATURAL AGAIN. THE TOXLINGS ALMOST HAD US. WE MUST UNDO WHAT THEY DID.

THE BIG CLEAN-UP

Angelina's words were heard around the world. The people of the world joined together to undo what had been done by the Toxlings. New rules were put in place to stop all pollution.

The biggest litter pick ever seen took place. They removed almost all the plastic litter from the oceans, towns and fields.

Only <u>renewable</u> energy sources were used after that. No more toxic nuclear waste and no more harmful gases from fossil fuels. Farmers used natural ways to keep pests off crops.

Nooooooooo! All that beautiful toxic waste is going to... waste!

Let's get out of here before they try to clean us up too!

EYES ON THE SKIES

That is how a young girl from a small town saved the world. If anyone is ever thinking about using something toxic, they just think, 'what would Angelina do?'

Angelina still watches the skies... just in case the Toxlings ever decide to return to Earth...